Quick Tricks
for Math

by Barbara F. Backer
illustrated by Marilynn G. Barr

To my grandchildren, Eric, Patrick, and Sydney,
who know all of the "quick tricks"
for bringing me joy.

Publisher: Roberta Suid
Design & Production: Standing Watch Productions
Cover Design: David Hale

Monday Morning Books
P.O. Box 1680
Palo Alto, CA 94302

E-mail us at: MMBooks@aol.com
Visit our Web site: www.mondaymorningbooks.com
Call us at: 1-800-255-6049

ISBN 1-57612-130-5

Printed in the United States of America
987654321

CONTENTS

INTRODUCTION

Math is all around us, offering opportunities for exploration and play. We count the petals on a flower and measure sunflowers. We estimate the number of doodle bugs under a rock, then we count to check our estimate. We collect acorns and divide them among friends, count family members to see how many cupcakes to buy, and compare sunflowers and marigolds to see how they are the same and how they are different. We count the number of days until Grandmom's visit, and we watch the numbers spin at the gasoline pump.

Children who have the opportunity to play with mathematical concepts grow into adults who enjoy math. Engineers, pharmacists, cooks, repair experts, taxi drivers— all of us use math every day in our work. We determine how long it takes to get to the airport, how to double a soup recipe, and how many days until pay day.

The math activities in this book are designed around children's interests. Preschool children learn best from hands-on experiences with learning materials. Therefore, the children will be encouraged to talk and be excited about the learning, and they will manipulate items to learn more about their mathematical qualities.

These "Quick Tricks" are meant to be enjoyed in a casual, informal setting and atmosphere. The emphasis is on having a mathematical experience, not in getting a right or wrong answer. If children come up with incorrect information, ask them how they figured that out. Perhaps you can suggest another way to explore. Remember that young children's brains have not finished developing. They are concrete thinkers and their way of looking at things is different from adults' methods. Through repetition of activities like those in this book, children build their knowledge and their problem-solving abilities.

Everyday materials and your enthusiasm are all that are needed for these games. Only a few activities require advance preparation.

Although you can use the activities in this book in any order, they are arranged with early chapters building skills needed for concepts in the later chapters. For example, a child who cannot recognize numerals cannot successfully play a game where he is to put five items in a bag labeled with the numeral 5.

Most chapters include a letter for parents. The letters inform parents of what their children are learning in school and offer a Quick Trick for continuing the learning at home. To keep the language simple, chapters alternate between referring to children as "he" or "she."

Quick Tricks for Math ©2001 Monday Morning Books, Inc.

COUNTING

As children begin learning math skills, they first learn to name the numbers. Many children can count by rote to a rather high number, but if you gather a dozen bottle caps and ask them to give you three, they are unable to do so.

The next stage in counting is realizing that the number names actually can tell you "how many." When children are beginning to learn this "meaningful counting," count just a few items at a time.

Children learn to count by participating in counting activities again and again. Take advantage of the opportunities to count.

Fuzzy Bears
A Quick Trick with Teddy Bears

Gather These Materials:
5 teddy bears

Where: anywhere

How: Place one bear in front of the children, starting at their left (your right). As you say the rhyme, have a child add a bear to the right of the previous bear each time another bear arrives. Show the children how to wiggle their fingers and fly them through the air to represent the bees. At the last line of the rhyme, gather the bears and whisk them away.

> One fuzzy teddy bear lived in the zoo,
> He called to another bear, then there were two.
> Two fuzzy teddy bears had cookies with their tea.
> They called to another bear, then there were three.
> Three fuzzy teddy bears liked to snort and roar.
> They called to another bear, then there were four.
> Four fuzzy teddy bears found honey in a hive.
> They called to another bear, then there were five.
> "Buzzzzz," said the bees. "You'd better not stay."
> And five fuzzy teddy bears quickly ran away.

Variation:
You can use precut paper shapes that look like bears.

This activity also helps children learn about:
adding by one.

Bracelets & Bangles
A Quick Trick with Pipe Cleaners

Gather These Materials:
pipe cleaners or chenille stems

Where: anywhere

How: Show the children how to bend a pipe cleaner into a circle and then twist the ends together to hold the shape. As they make bracelets, have them put the bracelets on one arm. Each time a child adds one, have her stop and count the bracelets. When she has more bracelets than she can count, have her move some bracelets to her other arm and then count separately the bracelets on each arm. When she again has too many to count, stop counting and just make bracelets for fun.

Variations:
Count only the red bracelets, the green ones, and so on. For extra counting practice, have the children count each other's bracelets.

This activity also helps children learn about:
number combinations when they see that two red bracelets plus one green bracelet are the same number as three blue bracelets. Also, two red bracelets and one green bracelet make the same total as two green bracelets and one red bracelet.

1, 2, 3, 4,
5, 6, 7, 8

Quick Tricks for Math ©2001 Monday Morning Books, Inc.

Pirate's Loot
A Quick Trick with Coins and Dominoes

Gather These Materials:
file folders
ruler
marker
coins
box decorated simply to look like a pirate's treasure chest
dominoes
pillowcase (or paper bag)

Where: at a table or on the floor

How: Cut a file folder along the fold, making two game boards.
On each, draw a square 7-and-1/2 inches (18.7 cm) on each side.
Divide the square into a grid of twenty-five 1-and-1/2-inch (3.7-cm)
squares. Each player uses one game board; make as many
as you need.

For each person playing, place 25 coins in the decorated box
and place the box in the center of the group.

Check the dominoes. Remove any that have more than eight
total dots on them. Place the remaining dominoes in the pillowcase. In turn, and without looking, each player removes a domino
from the pillowcase. She counts all of the dots on the domino then
counts out that number of coins and places them (pirate's loot) on
her grid. All players do the same, in turn. If a player draws a domino
with a blank on one side, she must empty her grid. The first player to
cover her grid wins.

Variation:
To add to the action for older players, if a player draws a domino
that is a "double," she takes half that amount from the pirate's
chest and the other half from one opponent's game board.

This activity also helps children learn about:
combining numbers (adding) as they count the dots on one side of
the domino and continue by counting (adding on) the dots on the
other side.

Five Hot Dogs
A Quick Trick with Crayons and Paper

Gather These Materials:
crayons
5 sheets of paper for each child

Where: at a table

How: Have your children draw a red hot dog on each of their sheets of paper. Have them color "yellow mustard" and "green relish" on each.

Spread five hot dog pictures in front of the group and together count the five hot dogs. Say the rhyme below and have a child remove one hot dog picture at each appropriate time.

Five Red Hot Dogs

Five red hot dogs with mustard galore.
Mother ate one, and then there were four.
Four red hot dogs with green relish you see.
Father ate one, and then there were three.
Three red hot dogs, yes, it's true.
Our dog ate one, and then there were two.
Two red hot dogs sitting in the sun.
Brother ate one, and then there was one.
One red hot dog on a soft brown bun.
I ate that hot dog and now there are none.

Variation:
Change the names in the rhyme, as desired. It's fun to substitute the names of children in the group.

This activity also helps children learn about:
subtracting by one.

Brother ate one, and then there was one.

Blowing Leaves
A Quick Trick with Precut Shapes

Gather These Materials:
5 precut paper leaf shapes
small pieces of tape
a twig

Where: anywhere

How: Place the twig in front of the children. Lightly tape five leaf shapes at various places on the twig. Together with the children, count the leaves; be sure that you touch a different leaf as you say each number name. Together, sing the song below. At the end of each verse, remove one leaf and count the remaining leaves.

Tune: "Baa, Baa, Black Sheep"

Five leaves are on the tree today,
Hanging on a branch so gray.
Blowing in the wind and sun,
Having lots of windy fun.
If one leaf should blow away, (remove one leaf)
Four leaves are on the tree today.

Four leaves are on the tree today....

(Continue song until no leaves are on the tree.)

Variation:
When the children become proficient with numbers through five, gradually add more leaves to the twig and more verses to the song.

This activity also helps children learn about:
subtracting by one as each leaf blows away.

> Five leaves are on the tree today.

Dear Parents,

We are practicing counting at school. Many children can count by rote (naming numbers in order) to a rather high number, like 10. However, if you gather a dozen bottle caps and ask them to give you three, they cannot do so. These children are learning number names. The next stage will be realizing that the number names actually can tell "how many."

You may have seen a child counting five objects. She touches two or more objects as she says a number name, or she may say the number names faster than she points. She may count to six or seven even though there are only five items. This is a normal part of counting development. To help your child grow through this stage, show her how she can <u>touch</u> each item and <u>say one number name only with each touch.</u>

While your child is beginning to learn this "meaningful counting," count just a few items at a time (two or three) and only move on to more when she is comfortably and correctly counting those few. Don't rush the process. Children learn to count by participating in counting activities again and again. Take advantage of the opportunities to count throughout the day. Help your child gather things to count and provide many chances to play with them. The "Quick Trick" below tells how you can practice counting skills at the grocery store.

A Quick Trick with Groceries

Gather These Materials:
shopping cart
groceries as you shop

Where: grocery store

How: Play counting games with your child in the grocery store. Let her count fresh produce as you put items in bags: three tomatoes, four red apples and two green apples, one bunch of celery. Count canned goods as you put them in the shopping cart. At the checkout stand, count items with green or red labels as you put them on the counter. Each time you visit the grocery store, let your child decide what you will focus on counting that day.

Quick Tricks for Math ©2001 Monday Morning Books, Inc.

NUMERAL FORMATION

Numerals are the written forms of numbers: 0, 1, 2, 3, 4, 5, 6, 7, 8, and 9. They signify their own numerical value and can be combined (123, 859) to signify other number values.

Children learn to form numerals through practice. The best practice begins by making large numerals on paper with paint or markers or by forming numerals of pliable materials like modeling dough. Keep it fun, and don't worry about whether children form the numerals correctly. Don't worry, either, about whether or not they know the names of or understand the value of the numerals they are making. Remember that the children are practicing.

Fuzzy Numbers
A Quick Trick with Pipe Cleaners

Gather These Materials:
calendars
pipe cleaners or chenille stems

Where: anywhere

How: Cut the pages from used monthly calendars. Give each child a page of numbers and some pipe cleaners. Encourage the children to study a numeral on their calendar page and then to bend a pipe cleaner to form that number. Encourage them to form more pipe cleaner numerals.

This activity also helps children learn about:
how two numerals placed side-by-side form another number. For example, the numerals 1 and 5 placed side-by-side make the number 15; a 2 and a 7 make the number 27.

Pretzel Numerals
A Quick Trick with Pretzels

Gather These Materials:
waxed paper
pretzel circles
pretzel sticks

Where: at a table

How: Begin by having the children wash their hands. Give each child a piece of waxed paper to use as a working surface. Put pretzels where the children can reach them.

Show the children how to fit two circle pretzels together to form the numeral 8. Encourage them to do the same, then tell them to eat their 8s.

Use pretzels to make other numerals. Make a 3 with two halves of a broken circle. Make 1, 4, and 7 with whole and broken pretzel sticks. Use sticks and circles to form 6 and 9. Use sticks and broken circles to form 2 and 5. Use a whole stick and a circle to form 10.

Variation:
Have children count the number of pretzel pieces they need for forming each numeral.

This activity also helps children learn about:
how several parts go together to make a whole.

Gluey Numerals
A Quick Trick with Index Cards and White Glue

Gather These Materials:
pencil
10 index cards for each child
white glue
white paper
spring-type clothespin or clip
old crayons with the paper peeled off

Where: at a table

How: Show your children an exciting way to make 10 numeral cards. For each child, do the following: On 10 index cards, lightly write large numerals 0-9, writing one numeral on each card. Have the child carefully trace over the pencil letters with white glue. Set the glue cards aside to dry for a day or two.

Cut pieces of paper the same size as the index cards. Show the children how to use the clothespins to clip a piece of paper on top of a numbered index card. Now have each child rub the side of a peeled crayon over his paper. The numeral on the card below will magically appear on the top sheet of paper. Have the child name the revealed numeral. If he cannot, name it for him.

Variation:
For added learning, after a child has made one of each numeral, let him put them in order from 0-9.

This activity also helps children learn about:
numeral recognition when they name the numerals they have produced.

Cool & Creamy
A Quick Trick with Yogurt

Gather These Materials:
spoons
unbreakable plates or cookie sheets
fruit-flavored yogurt

Where: at a table

How: Have your children wash their hands thoroughly. Let each
child spread a few spoonfuls of his favorite yogurt flavor on a plate
or cookie sheet. Encourage him to use his fingers to "write" numerals
in the yogurt and then spread the yogurt around on the plate to
erase the numerals. Continue as long as the children are interested.

Encourage the children to count items in the room (chairs, windows) and write those numerals in the yogurt. Suggest they write the
numerals that tell their ages. When they grow tired of the activity, let
them lick their fingers clean before they wash their hands.

Variation:
Print numerals on index cards. Display these on the table near the
children so they can copy the numerals by writing in the yogurt.

This activity also helps children learn about:
number names, when you talk about the numerals they are forming.

Quick Tricks for Math ©2001 Monday Morning Books, Inc.

Phone Numbers
A Quick Trick with Modeling Dough

Gather These Materials:
marker
sentence strips
clear, adhesive-backed paper
modeling dough

Where: at a table

How: For each child, do the following: With the marker, write the child's phone number on a sentence strip in 2-inch (5-cm) high numerals. Cover the sentence strip with clear, adhesive-backed paper. Encourage the child to look at the numerals and then use modeling dough to form each numeral. Have him put the modeling dough numerals in order below his sentence-strip model so that he forms his phone number. Say the numbers with him as he points to each numeral.

Variation:
Let the children swap phone number strips and form the numerals in a friend's phone number.

This activity also helps children learn about:
their phone numbers. As they form their own number again and again with modeling dough, eventually they will memorize it.

Toothsome Treat
A Quick Trick with Toothpaste

Gather These Materials:
plastic table cloth
cookie sheet, 1 for each child participating
waxed paper, a sheet for each child participating
toothpaste

Where: at a table

How: Cover the table with a plastic cloth. With each child, do the following: Cover a cookie sheet with waxed paper and put it on the table in front of the child. Squirt a 4-inch (10-cm) column of toothpaste onto the waxed paper. Have the child put his fingers into the paste and spread it on the paper. Show him how to drag his fingers through the paste to form "finger-paint" numerals. Talk about the numerals he is forming. Challenge the child to form the numeral that tells his age. Help him create numerals he sees in the room and those he can name. Add more toothpaste as necessary.

Variation:
Use hand lotion instead of toothpaste. <u>Caution:</u> Be certain the children keep their gooey hands away from their eyes and mouth.

This activity also helps children learn about:
number names.

Quick Tricks for Math ©2001 Monday Morning Books, Inc.

Dear Parents,

As we continue our math studies, we are focusing on numerals. These are the written forms of numbers: 0, 1, 2, 3, 4, 5, 6, 7, 8, and 9.

Children learn to form numerals through practice. The best practice begins by making large numerals on paper with paint or markers or drawing numerals outdoors in loose dirt or sand. Children also gain experience while making numerals of pliable materials like modeling dough, pretzel dough, and bread dough. Don't worry about whether your child forms the numerals correctly or about whether or not he knows the name of or understands the value of the numerals he is making. The activity below gives children a delicious way to practice making numerals.

A Quick Trick with Cookie Dough

Gather These Materials:
tablespoon
sugar cookie dough
(a tube of refrigerated dough is fine)
waxed paper
cookie sheet

Where: at a table

How: Place a rounded tablespoon of chilled sugar cookie dough on a piece of waxed paper in front of your child. Show him how to roll the dough into a "snake" and how to use the snake to form a numeral. Place the numeral on the cookie sheet. Continue until all the dough has been used. Help your child name the numerals he forms. Bake the cookies according to recipe directions. Watch carefully. You may need to bake for a longer or shorter amount of time. When the cookies cool, let him name the numerals that he eats.

NUMERAL RECOGNITION

Young children are beginning to realize that "squiggles on a page" have meaning. They usually learn to identify letters before numerals because the letters are more important to them. They recognize the letters in their own names and the names of those in their family.

Often the first numeral a child recognizes is the one that identifies her own age. For example, the four-year-old's favorite numeral is usually four. Take advantage of this interest by pointing out to each child the numeral that shows her age. Then point out other significant numerals. Over time, the children will become familiar with all of the numerals, but it will be a long time before they understand the difference between a "number" and a "letter."

Microwave Numerals
A Quick Trick with the Microwave Oven

Gather These Materials:
microwave oven

Where: in front of a microwave oven

How: With a small group of children (so everyone can see) have a child call out a number. Enter that number so it shows on the oven's display. After everyone has a chance to look at the number, clear the display and begin again.

When you are cooking with the microwave oven, call out the numbers as you enter them.

Variation:
When you are using a microwave oven, have your children watch the numerals change as the oven "counts down" the remaining cooking time.

This activity also helps children learn about:
counting down.

Quick Tricks for Math ©2001 Monday Morning Books, Inc.

Number LOTTO
A Quick Trick with Index Cards and Bottle Caps

Gather These Materials:
large index cards
small index cards
marker
scissors
bottle caps for LOTTO markers, 6 for each child

Where: at a table

How: The adult does this: Using six or more of the large index cards, make several LOTTO cards by dividing each card into six sections. Write a numeral (0-9) in each section, making each card different. (Some numerals will be the same on two or more cards, but they can be placed in different locations.) Cut five small index cards in half. Write one numeral 0-9 on each card.

With a small group of players, have all children but one select a large LOTTO card and six bottle caps to place on the table in front of themselves. Have the remaining child stack the small cards face-down. Ask her to turn over a small card, identify the numeral, and show the numeral card to the group. (If she cannot identify the numeral, the group can help her.) If a child has that numeral on her large LOTTO card, she covers it with a bottle cap. The caller continues turning over small cards and showing and naming the numeral. When all numerals on a player's card are covered, that player shouts, "LOTTO!" Continue play until every player has called out "LOTTO!" Have players change cards, choose another child to be the caller, and repeat the game.

Variation:
For two or three children, each selects a LOTTO card. In turn, they turn over a small numeral card. If that numeral is on their own LOTTO card, they cover it with a bottle cap. At the end of each turn, the small card is returned to the bottom of the face-down deck. Children continue turning over the numeral cards until one can shout "LOTTO!"

This activity also helps children learn about:
focusing their attention.

Fishing for Numerals
A Quick Trick with Paper and Magnets

Gather These Materials:
24-inch (60-cm) string
18-inch-long (45-cm) dowel
small magnet
markers
construction paper
child-safe scissors
paper clips

Where: the floor

How: The adult does the following: Tie one end of the string onto the end of the dowel to make a fishing rod. Tie or tape the magnet to the string's other end. With your children, draw and cut out construction paper fish shapes. Using the numerals 0-9, write one numeral on each fish and place a paper clip on the nose of each fish.

Spread the fish face-up in a 3-foot (1-m) square area on the floor. In turn, have your children try to catch a fish by holding the fishing pole and dangling the fishing line directly over the fish. When the magnet touches a fish's paper clip, the fish will stick to the magnet, and the child can "bring in" the fish. Have her look at the numeral and name it if she can. If she cannot, name it for her. Children keep the fish they catch until the end of the game. At that time they "throw back" their fish, placing them in the fishing area.

Variation:
Make several fish for each numeral. After the children catch the fish, have them put matching numerals together.

This activity also helps children learn about:
properties of magnets.

Sticky Numbers
A Quick Trick with Magnetic Numerals

Gather These Materials:
magnetic plastic numerals (available at dollar stores and toy stores)

Where: on any magnetic surface—refrigerator, file cabinet, cookie sheets

How: Spread out the magnetic numerals randomly on the magnetic surface. Encourage a child to find each numeral in order and to replace the numerals on the magnetic surface in order as together you slowly count out loud. If she needs help, tell her where to look. ("The 3 is green. It's beside a yellow 2." "The 5 is under a purple 9.") Call other children to see the ordered numerals and praise the child for her accomplishment.

Variation:
Put the magnetic numerals on the floor. Call out numbers to a child and have her find those magnetic numerals and put them on the magnetic surface.

This activity also helps children learn about:
sequence of numbers.

Scan the Ads
A Quick Trick with Grocery Ads

Gather These Materials:
markers of various colors
index cards, 1 set of 10 for each child
grocery ads, 1 page for each child

Where: at a table

How: For each child, use the marker to write one numeral (0-9) on each of 10 index cards. Give each child a set of cards and one page of a grocery ad. (Giving too many pages can confuse or distract a child.) Let her choose an index card from her set, look at the numeral, and find that numeral in the ad. Have her circle each matching numeral that she finds. Help her state the numeral's name. Have her choose a different-colored marker, select a new numeral card, and repeat the activity.

Variation:
Have your children look first at their advertising page and circle a numeral. Now have them look through their index cards until they find the matching numeral.

This activity also helps children learn about:
what's in newspapers.

Reading Coupons
A Quick Trick with Grocery Coupons

Gather These Materials:
child-safe scissors
Sunday newspaper advertising sections and other colorful advertising that contains grocery coupons; ask parents to send these in

Where: at a table

How: Encourage each child to cut out or tear out the grocery coupons. Let each child show you and the group a coupon, and let her tell you the numerals she sees on it. Help her identify any numerals she doesn't know. Beginners will not know two-digit numerals like 25. Let them label these by each digit—a "two" and a "five."

Variation:
Name a numeral and have each child find it in her grocery ads.

This activity also helps children learn about:
cutting skills.

Watch the Clock
A Quick Trick with a VCR

Gather These Materials:
VCR with a lighted display

Where: in front of the VCR where everyone can see the display

How: Set your VCR clock to the correct time. In small groups (so everyone can see), show your children how the numerals on the VCR clock display change every minute. Help them discover that the numbers go up by one each time they change. Say the numbers together as they change. See if the children can predict what number will come next. For younger children, just focus on the last digit in the display (numerals from 0-9). For older children, focus on the numbers that stand for the minutes in the time: (00-59).

Variation:
For children who are having success writing numerals, give them markers and paper and challenge them to copy the changing numerals on the VCR display.

This activity also helps children learn about:
the passage of time.

Dear Parents,

Young children are beginning to realize that "squiggles on a page" have meaning. They usually learn to identify letters before numerals because the letters are more important to them. They recognize the letters in their own names and the names of those in their family.

Often the first numeral a child recognizes is the one that identifies her own age. For example, the four-year-old's favorite numeral is usually four. Take advantage of this interest by pointing out to your child the numeral that shows her age. Then point out other significant numerals—her sister's age, the number of people in your family, the number of pets you have, the number of windows in her bedroom. Over time, she will become familiar with all of the numerals, but it will be a long time before she understands the difference between a "number" and a "letter."

The "Quick Trick" below gives your child a sweet reason to remember special numerals, and it provides warm family memories as well.

A Quick Trick with Pancake Batter

Gather These Materials:
pancake ingredients
mixing bowl and utensils
frying pan and utensils
plate and fork

Where: in the kitchen

How: Have your child help you mix up a simple pancake batter. Have her name a numeral for you to prepare. Pour the batter in a thin stream into the pan forming the numeral backwards in the pan. Flip the pancake when the first side is cooked, and you'll see the requested numeral. Put the fully cooked pancake on a plate for your child. If your child is unable to name a numeral for you to prepare, make one that is meaningful to her and tell her what it is. For example, "This is a four. You are four years old."

MATCHING NUMERALS TO NUMBER VALUES

Many children can count to 10 or further, but if you ask them to put four blocks in a basket, they are unable to do so. They know how to name numbers in order, but they don't understand that these numbers represent a specific value. Likewise, some children can name some of the numerals 0-9 when they see them, but they don't understand that each numeral represents a specific value. However, when a child puts three caps on an index card with the numeral 3, he is developing the ability to match numerals and number values.

Count & Eat
A Quick Trick with Snack Foods

Gather These Materials:
index cards
marker
large bowl
small snack foods
small paper cups

Where: at a table

How: Make number cards by writing a numeral (0-9) on each card. With the marker make a matching number of dots on each card—four dots on the four card, six dots on the six card. Provide a snack by placing separate bowls of small snack foods (raisins, pretzels, or whatever you have available) on the table where children can reach it. In front of each bowl, place a number card. Have the children wash their hands. Have each child, in turn, take a paper cup. Tell the children that they may take from each bowl the number of items indicated by its number card. They place these items in their paper cups, and when they've taken all they are allowed, they eat the contents for their snack.

Variation:
Offer a mixed combination of three small snack foods and provide a tablespoon. Also provide a small paper plate for each child and a variety of number cards. Let each child serve himself a tablespoon of mix. Have him sort the snack items into three separate piles and find number cards that show how many of each item he has.

This activity also helps children learn about:
taking turns.

Number Book
A Quick Trick with Paper and Rubber Stamps

Gather These Materials:
plain paper, 5 sheets for each child
construction paper, 2 sheets for each child
stapler and staples
marker
rubber stamps
stamp pad(s)

Where: at a table

How: For each child, make a book by gathering together and stacking five pieces of plain paper. Put a piece of construction paper on the top of the stack of papers and another on the bottom for book covers. Staple the stack of papers on the left-hand side. On each page of the book, write a numeral, in order, beginning with 1 on the first page and ending with 5 on the last. Also use alphabet letters to write each numeral's name.

Show your children how to use the rubber stamp. Have them open their books to the first page and stamp one time on that page. Have them stamp twice on page 2, three times on page 3, and so on. When they are finished, encourage them to read the books to each other and to their families.

Variation:
Make a similar book using stickers instead of the rubber stamps.

This activity also helps children learn about:
numeral recognition.

Counting Fingerprints
A Quick Trick with Fingerprints

Gather These Materials:
paper
marker
stamp pads
wet and dry paper towels

Where: at a table

How: For each child, fold a piece of paper in fourths, then unfold it. Write a numeral (1-4) in each of the paper's four sections. Put out stamp pads and show children how to press a fingertip onto the pad and then onto one section of their paper, leaving a fingerprint on the paper. Now have children make fingerprints in each section of their papers, matching the number of prints to the numeral in the section. Repeat the activity as long as they are interested. Provide paper towels for cleaning fingers.

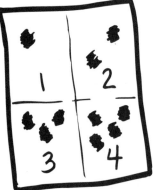

Variation:
Have the children make fingerprints first, then count the prints in each section and, if they are able, write the correct numeral in each section. (If a child cannot write the numerals, do so for him after he counts the prints.)

This activity also helps children learn about:
counting.

Egg-ceptional Math
A Quick Trick with Plastic Eggs and an Egg Carton

Gather These Materials:
6 plastic snap-apart eggs
permanent marker
cardboard egg carton cut in half so each half has 6 sections
small bowl
21 small buttons or other small items

Where: at a table or on the floor

How: Gather the plastic snap-apart eggs. Close each of them. Using a permanent marker, write one number from 1 to 6 on each egg. Put the eggs in the six-section egg carton.

Fill a small bowl with buttons or other small objects. Have a child count out the objects to fill each egg, matching the number of items with the number on the egg. Have him open the egg, put in the correct number of items, and return the filled eggs to the carton. The next child to play the game cracks open each egg and spills the contents into the bowl before beginning to count.

Variation:
As your children's skills increase, add more eggs with larger numbers, until you have 12.

This activity also helps children learn about:
hand-eye coordination.

Get a Handle on Numbers
A Quick Trick with Laundry-Detergent Scoops and Buttons

Gather These Materials:
laundry-detergent scoops
scissors
index cards or white paper
pen
small plastic bowl
15 paper clips

Where: at a table

How: Gather six laundry-detergent scoops. Cut index cards or white paper to fit into the bottom of each scoop. Write one numeral, 0-5, on each index card, then glue one card into each scoop. Give a child a plastic bowl containing 15 paper clips. Have him look at the numeral in a scoop. Then have him count out the correct (matching) number of paper clips and place them in the scoop. Continue until all scoops have the proper number of clips. Have the child pour the clips back into the bowl when he is through with the game.

Variation:
Substitute other small items (buttons, nuts, acorns) for the paper clips. Each time you change items, the children will think it is a new game.

This activity also helps children learn about:
cleaning up when they have finished playing with a game.

Number-O's

A Quick Trick with Pipe Cleaners and O-Shaped Cereal

Gather These Materials:
masking tape (any width)
6 pipe cleaners
pen
O-shaped cereal

Where: at a table

How: The adult does the following: Tear off six 2-inch (5-cm) pieces of masking tape. Place one end of a pipe cleaner at the center of a piece of tape, on the sticky side. Fold the tape in half with the sticky sides touching. Press the tape together forming a "flag" of tape at the end of the pipe cleaner. Place the pipe cleaner on a table so the "flag" is on the left side of the pipe cleaner and is pointing to- ward your body. Write the numeral 0 on the flag. Continue with the remaining pipe cleaners, numbering them from 1 to 5.

Have the children wash their hands. Give the pipe cleaners and a bowl of O-shaped cereal to a child. Have him string cereal pieces on each pipe cleaner so that each pipe cleaner has the appropri- ate number of cereal pieces—none on the 0 pipe cleaner, four on the 4 pipe cleaner, and so on. He can fold over the right end of the pipe cleaner to keep the cereal pieces from sliding off. Clean up is fun—the child eats the cereal pieces as he removes them.

Make several sets of pipe cleaner "flags" so that several children can enjoy the activity together. Use a different color of pipe cleaner for each set of flags.

Variation:
Children learn about combining numbers when you make the game with fruit-colored O-shaped cereals and colored pens. On the tape flags, write simple addition problems. Write the first and second addend with different colored pens. Have children count out two colors of cereal, using a different color for each addend on the flag.

This activity also helps children learn about:
hand-eye coordination.

Find the Numeral
A Quick Trick with Index Cards and Stickers

Gather These Materials:
a variety of <u>small</u> stickers
10 large index cards
LOTTO cards from "Number LOTTO" (see page 19)
bottle caps

Where: at a table or on the floor

How: Have children help you make a set of cards for this game by putting stickers on each of 10 index cards (no sticker on one card, one sticker on one card, two stickers on another card, and so on up to nine stickers on the last card).

Have each of several children select a LOTTO card from "Number LOTTO" (page 19) to use as a game card. Mix up the sticker index cards and then place them face-down in a pile. Have the players take turns drawing from the pile. If the child who drew the card has on his LOTTO card a numeral that matches the number of stickers, he places a bottle cap over that numeral on his game card. When the children reach the bottom of the sticker pile, have them remix the cards and make a new, face-down drawing pile. Continue play until one player covers all of his numerals and shouts "LOTTO!" Players can then clear off their cards and repeat the game, with the last game's winner drawing first from the new pile.

Variation:
Have children use raisins or small crackers to cover the numerals on the game cards. At the end of each game, let them eat the markers they've used.

This activity also helps children learn about:
taking turns.

Leprechaun's Gold
A Quick Trick with Index Cards and Dried Beans

Gather These Materials:
1 pound (500 gr) dried beans
gold spray paint, for adult use only
marker
index cards
plastic dish or container, 1 for each player

Where: at a table or on the floor

How: The adult does the following: When children are not around and the weather is calm, not windy, spread the beans on newspapers outside. Cover all nearby surfaces with newspaper. Spray paint the beans. Allow them to dry and then turn them over and spray again. These golden beans become the leprechaun's gold.

With a marker, write numerals on the index cards. Make several cards for each number. Use numerals that your children are able to read and understand.

With children, do the following: Place the cards face down in a stack where everyone can reach them. Scatter the "gold" in front of the players where all can reach it. Have each player select a plastic dish for his "pot of gold."

Each player in turn pulls a card, reads the numeral (with help, if necessary), and then counts out and puts that many "pieces of gold" into his pot. Continue play until all the gold pieces are taken. Don't worry about "winners." The object of this activity is to associate number values with written numerals and to practice counting. Be available to help the children read the numerals and count out their gold.

Variation:
Use coins, bottle caps, or foil-wrapped candies instead of painted beans.

This activity also helps children learn about:
small motor coordination.

Dear Parents,

When children begin to understand that written numbers (numerals) represent a specific value, they take a major step in learning math. They are combining a reading skill with a math skill and are also beginning to use abstract thinking skills. In abstract thinking, a symbol (the written number, 3) stands for a concrete thing (a specific amount of things, three beans).

Children first learn to name (read) a few of the numerals when they see them. The understanding that each numeral represents a specific value soon follows.

When your child puts three bottle caps on an index card with the numeral 3, he is developing and practicing the ability to match numerals and number value. Take advantage of pointing out numerals 0-9 in your child's environment and asking him to show you "that many" of something: fingers, buttons, shoes. Try the "Quick Trick," below, for another way of matching numerals and number value

A Quick Trick with Envelopes and Paper Clips

Gather These Materials:
6 used (or new) envelopes
marker
15 paper clips
plastic dish (recycled margarine tub or deli container works fine)

Where: at a table or on the floor

How: On the back of each envelope, write a different numeral, 0-5. Show your child how to slide a paper clip onto the edge of an envelope. Help him read the numerals on each envelope. Have him count out that number of paper clips and slide them onto the envelope. Have him continue with the remaining envelopes. As his skill with numbers improves, add more paper clips and more envelopes with higher numerals on them.

ORDINAL NUMBERS

> Ordinal numbers tell the order of things: first, second, third, and so on. Take advantage of opportunities to use these words with the children in everyday situations.

Pet Parade
A Quick Trick with Stuffed Animals

Gather These Materials:
a small variety of stuffed animals

Where: on the floor

How: Have each child bring a stuffed animal to the group. Have the children arrange their animals in a row. With the children, sing the following song, substituting the correct ordinal number and animal name for the underlined words.

The Pet Parade
Tune: "The Farmer in the Dell"

The <u>tiger cub</u> is <u>first</u>,
The <u>tiger cub</u> is <u>first</u>,
Come see the pet parade,
The <u>tiger cub</u> is <u>first</u>.

The <u>panda bear</u> is <u>second</u>,
The <u>panda bear</u> is <u>second</u>,
Come see the pet parade,
The <u>panda bear</u> is <u>second</u>.

Final Verse:

They all marched around.
They all marched around.
They marched in the pet parade.
They all marched around.

> The tiger cub is first,

Variation:
Sing about the order of the children when they line up to go outdoors.

This activity also helps children learn about:
animal names.

Five Little Kittens
A Quick Trick with Stickers and Index Cards

Gather These Materials:
5 jar lids or large bottle tops
5 cat stickers
1 stuffed animal dog

Where: on the floor

How: Affix one sticker to each jar lid. Spread the lids, cat-side-up, on the floor in front of the children. Have one child hold the dog and sit near the cat lids. Beginning at the children's left, count the cats with them. Recite the following poem touching each cat in turn, beginning on the children's left. At the next-to-last line, have the child with the dog move it toward the cats and say, "Bow, wow, wow." As she "barks," sweep the cats up in your hand and "hide" them behind you.

> Five little kittens were playing on the floor.
> The <u>first</u> one said, "Let's hide behind the door."
> The <u>second</u> one said, "Let's play in the hall."
> The <u>third</u> one said, "I want to play ball."
> The <u>fourth</u> one said, "Let's scamper down the stairs."
> The <u>fifth</u> one said, "Let's play beneath the chairs."
> Along came a puppy dog—"Bow, wow, wow."
> Five little kittens are hiding right now.

Repeat with other children having a turn to hold the dog.

Variation:
Instead of stickers and jar lids, use commercially available precut paper shapes. Or, if you have an abundance of stuffed animals, use them to act out the rhyme.

This activity also helps children learn about:
dramatization.

Bow, wow, wow

Remember When?
A Quick Trick with Paper

Gather These Materials:
plain paper, 1 sheet for each child
(cut a standard sheet of paper in half, lengthwise)
fine-line markers

Where: at a table

How: Accordion-fold each paper into an accordion book of four pages. With your children, review the activities of their day. Focus on one particular event, like a cooking activity or a field trip. Talk with the children about what they did first, second, third, and last. On their books' first pages, have the children draw a picture of something they did <u>first</u>. On the second page, they can draw something they did <u>second</u>, and so on. Have children dictate to you the words to put on each page. Be certain to include the ordinal numbers <u>first</u>, <u>second</u>, and so on. Title the book: "Remember When ?..." ("Remember When We Went to the Museum?" "Remember When We Made Pizza?")

Read the books to the group, then encourage the children to read their books to each other. Have the children take their books home to read to their families.

Variation:
Use wrapping paper or construction paper to make a simple envelope/pocket for the book.

This activity also helps children learn about:
the passage of time when they read the book and remember what happened.

Story Time Fun
A Quick Trick with Storybooks

Gather These Materials:
a child's storybook

Where: anywhere

How: After you read a story to the children, take a few minutes to review the story informally. Talk about what happened <u>first</u>, <u>second</u>, <u>third</u>, and so on. Don't expect them to remember every detail. For example, in the story "Three Billy Goats Gruff," they might remember who crossed the troll's bridge first, second, and third without remembering that the goats set out one day to eat grass on the hillside. Focus on the order of the story, and only cover two, three, or four events.

Variation:
Recall a group memory, perhaps a special visitor or a day you found flowers blooming in your play yard. Talk about what happened first, second, third, and so on.

This activity also helps children learn about:
memory skills.

First, one billy goat crossed the bridge.

Quick Tricks for Math ©2001 Monday Morning Books, Inc.

Dear Parents,

We have been learning about <u>ordinal</u> numbers. We use these many times a day. They're the numbers that tell us the order of things: first, second, third, and so on.

Take advantage of opportunities to use these words with your child in everyday situations. Describe things that are about to happen by using ordinal numbers: "First, we'll put on our jackets. Second, we'll get in the car. Third, we'll go to Grandma's house." Your child will learn how to use ordinal numbers just from hearing you use them correctly.

You can also ask your child questions using ordinal numbers in the questions. At bath time, you might ask your child what you need to do first and second: "First, we'll put water in the tub. Second, I'll get in the tub."

In the "Quick Trick" below, your child will be the star of a game that helps her use ordinal numbers.

A Quick Trick with Photographs

Gather These Materials:
camera (preparation)
film (preparation)
photos (activity)

Where: in the bedroom and bathroom (preparation)
anyplace you can spread out the photos (activity)

How: Take several photos of your child while she prepares for bed one evening. Capture her in various parts of her getting-ready-for-bed routine: taking off shoes, wrapping up in a towel after the bath, wearing pajamas, brushing teeth, getting in bed, sleeping.

Show the pictures to your child. Have her find the photo that shows what happened first the evening you took the photos. Have her find the photo that shows what happened second, then third, fourth, fifth, and sixth. Have her place these in order (working from left to right) from first to last. Talk with her about the photos, using the words "first," "second," "third," and so on while you talk.

MATCHING AND COMPARING

Children use visual discrimination and their powers of observation when they compare items or groups of items to see if they are the same or different. When you present activities with this skill, start with only two categories of items. Have children divide a collection of two sizes of paper clips into two groups—large and small. Give them blocks of two colors to group by color. Describe things around you as hard or soft, hot or cold, rough or smooth, fast or slow.

Later, give children a collection of shoes or mittens and let them put these in pairs. Use the words "match," "same," and "different" as you explore items in your environment.

All of the Animals
A Quick Trick with Stuffed Animals

Gather These Materials:
stuffed animals

Where: anywhere

How: Have the children gather together a variety of stuffed animals (ones they have brought from home or ones you have on hand).

Talk about the ways the animals are the same. Do they all have eyes? Legs? Tails? Are they all new? Are they all brown? When you come to a category where they differ, have the children look carefully at each animal for that characteristic and divide them into two groups—those that have the quality and those that don't. Discuss the results. Which group has more?

Variation:
Encourage children to think of other ways to compare the items.

This activity also helps children learn about:
observing for detail.

> Five have no stripes. One has stripes.

Quick Tricks for Math ©2001 Monday Morning Books, Inc.

Calendar Matching
A Quick Trick with a Calendar

Gather These Materials:
scissors
discarded wall calendar with large pages

Where: at a table

How: Remove a page from a calendar. Use a month that has 30 days: April, June, September, November. (Do <u>not</u> use number squares that are divided diagonally to serve two dates.) Cut along the calendar lines to make 30 number cards, 1-30.

Remove another 30-day page and leave it uncut. Have a child mix up the number cards and then draw them, one at a time, from a face-down pile. Have him look at the number he draws, find its match on the intact calendar page, and put the number card on that match. Help him identify each number he selects. Make several of these games so that several children can play together. They'll learn by helping each other find the matching numbers.

Variation:
For younger children, use only the first two weeks of a calendar page and the matching numbers needed to play the game.

This activity also helps children learn about:
the names of numbers from 1 to 30.

Circles Around
A Quick Trick with Jar Lids

Gather These Materials:
damp sponge
pie plate or leak-proof paper plate
tempera paint
2 jar lids, 1 small and 1 large
plain paper, 1 sheet for each child

Where: at a table

How: Have a child put the sponge in the pie plate. Help him pour a little tempera paint onto the sponge and poke the sponge so it absorbs the paint. In turn, have each child do the following: Press the open side of a jar lid into the paint on the sponge and then press the paint-covered rim onto paper. Repeat a few times. Repeat the activity with the other jar lid. Discuss together (and with other children) the imprints left on the paper. Encourage children to use the words *big, little, large, small, larger, smaller.*

Variation:
Repeat the activity with each child using a sheet of paper with a vertical line dividing the paper into two columns. On the top of one column, write "large" and on the other column, write "small." Have each child make circle printings in the correct columns.

This activity also helps children learn about:
different words for describing an item's size.

Quick Tricks for Math ©2001 Monday Morning Books, Inc.

Bowling for Math
A Quick Trick with Soda Bottles

Gather These Materials:
10 empty plastic soda bottles, 1-quart (1-liter) or 2-quart (2-liter) size
sand or dirt
chalk
a large ball

Where: outside or in a large indoor space

How: Fill each soda bottle one-third full with sand or dirt. If you don't have access to these materials, substitute rice, salt, or cornmeal.

Set up the bottles in a bowling pin configuration. Use chalk to mark off the foul line (beyond which the bowler should not go when releasing the ball).

Do the following with a small group of children: In turn, have your children roll the ball at the bottles. Count the number of pins a bowler knocks down and remove them. Have him roll the ball at the pins one more time and count the pins he knocked down. Put the knocked-down pins together to one side, so he can count and record the total of knocked-down pins.

All players roll the ball twice on their turns, and they count and record the total number of downed pins as their score for that round. Players compare their scores, and the player with the highest score goes first in the next round.

Variation:
Some children cannot count and record their scores. After you move their knocked-down pins to the side, have them put one stone or button beside each of these pins. Leave the stones in place while the next player bowls. Have that player do the same with his knocked-down pins. Now the children can line up their stones side-by-side to see who knocked down more pins.

This activity also helps children learn about:
counting.

The Yellow Pages
A Quick Trick with the Telephone Book and a Toy Phone

Gather These Materials:
telephone book
discarded telephone or toy telephone

Where: anywhere

How: Show your children how to find familiar logos in the advertising section of the telephone book. You might want to mark some pages (restaurants, grocery stores, toy stores, book stores, florists) with tape flags so the children can easily find them.

Show them how to find telephone numbers in the book and how to push those same number buttons on the discarded phone. The children can pretend to talk to people at the numbers they called.

Be certain the children understand that they are to call these numbers only on nonworking or toy telephones.

Variation:
Help the children find telephone numbers in newspaper and magazine advertising and in catalogs. Let them pretend to call those places to order items.

This activity also helps children learn about:
finding information in printed resources.

Hello. Is this the New Golden Dragon restaurant?

Matching Magnets
A Quick Trick with Frozen Juice Lids

Gather These Materials:
12 same-size lids from cans of frozen juice
permanent marker
magnetic tape

Where: on the refrigerator

How: Collect 12 same-size lids from cans of frozen juice. Divide the lids into six groups of two lids each. Using numerals 0-5, use the marker to make matching lids by writing the same numeral on both caps in a pair of juice lids. You should have pairs of each numeral from 0 to 5. Cut twelve 1-inch (2.5-cm) pieces of magnetic tape (available at hardware or craft stores). Peel the backing off the tape and press a strip of the magnetic tape onto the back of each juice lid. Place the lids at random on the side of a file cabinet or on a refrigerator door. Encourage children to find the ones that are the same and put them together.

Variation:
When your children have mastered matching the numerals 0-5, add more lids and let children match numerals up to 10.

This activity also helps children learn about:
recognizing numerals.

Feel This
A Quick Trick with Construction Toys

Gather These Materials:
an assortment of construction toys: 2 identical pieces of each toy
(2 Bristle Blocks, 2 Tinker Toys, 2 LEGOs, 2 wooden blocks,
2 snap-together blocks, and so on)
a dark-colored pillowcase

Where: anywhere

How: Place several pairs of assorted construction toys in a dark-colored pillowcase. Tell the children that they will be able to select matching items from the pillowcase without using their eyes. In turn, invite each child to reach into the pillowcase, without looking, and to find two matching items. Have the child remove the items from the pillowcase to see if they are, indeed, identical. Return the items to the pillowcase, shake it to mix the items, then begin again.

Variation:
Select several pairs of blocks from just one kind of building toy (all LEGOs or all wooden blocks, for example). Each pair should be a different size or shape from all of the other pairs. Proceed as above having children feel the materials to find a pair.

This activity also helps children learn about:
focusing their attention.

Quick Tricks for Math ©2001 Monday Morning Books, Inc.

How Are We Alike?
A Quick Trick with Friends

Gather These Materials:
a group of children

Where: anywhere

How: Select two children to stand in front of the group. Ask the group to find some way that these two children are alike. (They both have eyes. Both are boys. Both are wearing shoes.) Have each of the first children choose another child to take his place, then begin the activity again. Continue playing until every child has had a turn to be in front of the group.

Variation:
Repeat the activity having three or four children stand in front of the group. Challenge the group to find ways that all of the children are the same.

This activity also helps children learn about:
paying attention to detail.

Which Is Bigger?
A Quick Trick with Ribbon

Gather These Materials:
scissors
ribbon

Where: anywhere

How: With each child, measure out a length of ribbon to match the child's height. Encourage the child to use the ribbon to measure objects in the room such as a table, chair, stuffed animal, building block, or door. Ask him if the item is bigger than the ribbon, smaller, or the same size. If the item is smaller, help him realize that he is bigger than the item. If the item is larger than the ribbon, help him understand that he is smaller than the item. If the ribbon and the item are the same size, help him understand that he and the item are the same size where it was measured. (He may be taller than the table's height, but not longer than its width.)

Variation:
Use the ribbon to measure outdoor items as well. Have a child wrap the ribbon around a tree trunk. Is the tree smaller around than he is tall? Larger? Is he smaller than a lawn chair?
Smaller than a tricycle wheel?

This activity also helps children learn about:
measurement.

I am bigger than this chair!

Quick Tricks for Math ©2001 Monday Morning Books, Inc.

Dear Parents,

We have been practicing the math concepts of matching and comparing. We've been learning to use the terms *more, less, larger, smaller, taller, shorter, greater than, less than,* and *the same.*

Your child has many opportunities to practice matching and comparing skills at home. It's best to start with only two categories. For example, have your child help you divide grocery items into two groups: those that go in the refrigerator and those that do not.

Use comparative words in your everyday language. Describe things around you as hard or soft, hot or cold, wet or dry.

The "Quick Trick" below gives your child ample opportunities to practice matching skills. The game is a lot of fun, so gather the family and play along with your child.

A Quick Trick with Playing Cards

Gather These Materials:
deck of playing cards

Where: at a table or on the floor

How: The object of this game is to be the person left holding the queen at the end of the game.

Sort through a deck of playing cards and remove all aces, kings, jacks, and three of the queens. For three or more players, use two decks of sorted cards and one queen. Deal five cards to each player and place remaining cards (the dealing stack) where all players can reach it. Remind players to avoid showing their cards to each other.

The first player draws a card from any other player. If the card he drew matches one in his own hand, he places that pair in front of himself and draws another card.

The child whose card was drawn now replaces that card with one from the dealing stack. (If it matches a card in his hand, he places the pair in front of himself.) Now he draws a card from any other player, and play continues as above.

Play continues until only one player, the winner, is left with a card, the queen. The winner then goes first in the next game.

SORTING AND CLASSIFYING

While the children were matching and comparing items, they were also building skills used in sorting and classifying. They have already been sorting items as they've put them into categories that you suggested: hot/cold, large/small.

The children will continue sorting with activities in this chapter and then move on to classification, putting items into groups using criteria that they select.

Remember that young children's thinking is concrete and they won't necessarily think of characteristics that would occur to you. Always ask children their reasons for grouping items together.

Nuts, Nuts, Nuts
A Quick Trick with Mixed Nuts

Gather These Materials:
a bag of mixed nuts in their shells
1 large bowl or basket
a variety of small baskets or plastic bowls
2 or 3 kitchen tongs

Where: at a table or on the floor

How: Purchase a bag of mixed, unshelled nuts. Pour the contents into a basket or bowl and put them where the children can explore them.

After the children have explored the nuts, add a variety of small baskets and the kitchen tongs to the basket of nuts. Challenge the children to use the tongs to transfer the nuts to the small containers. Encourage them to put all the nuts of one kind into one each small basket. When they finish, have them spill the nuts back into the large basket and mix them up so they'll be ready for the next group of players. <u>Caution:</u> Do not use this activity if any child in your group has a nut allergy.

Variation:
Have the children count the nuts in any two of the sorted groups and compare to see which group has more.

This activity also helps children learn about:
small muscle control.

Cap's Off
A Quick Trick with Bottle Caps

Gather These Materials:
assorted bottle caps

Where: at a table or on the floor

How: Collect a variety of bottle caps. Try to include a variety of colors and sizes. Some may have writing on them and others may be plain. Spread the bottle caps and look at them with a small group of children. Tell them you will take turns finding three bottle caps that are the same in some way. Ask a child to find three bottle caps that are the same and put them together. Have her tell you how the caps are the same. (Many children begin by dividing the caps by color.) Let the other children, in turn, gather three bottle caps and tell how theirs are the same. When it is your turn, try to think of a similarity that has not yet been mentioned (size, with writing, without writing, plastic, metal). Continue taking turns until a child cannot find three similar caps. If no one can find three, mix up all of the caps and begin again.

In time, the children will begin dividing items while focusing on two attributes at a time: "These bottle caps are small and red." "Some of these blue bottle caps have writing on them."

Variation:
This activity can be done with many different everyday materials—buttons, assorted screws, bread tags, discarded caps from dried-up markers, scraps of a variety of papers, fabric scraps.

This activity also helps children learn about:
paying attention to details.

These three caps have no writing on them.

Marvelous Marbles
A Quick Trick with Marbles

Gather These Materials:
a variety of marbles

Where: on the floor

How: Provide a variety of marbles for the children; the more they have, the better. In small groups, encourage them to explore the marbles. In time, they will begin to notice similarities and differences in the marbles and will arrange them according to color or size. Invite them to talk about what they are doing. Encourage them to count the marbles in each group. If they rearrange the marbles, encourage them to explain why and how they have changed them (their new categories for classifying).

Variation:
Provide several unbreakable containers with the marbles. The children will have fun pouring the marbles back and forth. Try boxes, plastic measuring cups, cookie tins, and plastic jars or food containers.

This activity also helps children learn about:
counting.

Quick Tricks for Math ©2001 Monday Morning Books, Inc.

Safe or Dangerous?
A Quick Trick with Magazine Pictures

Gather These Materials:
adult scissors, for adult use only
child-safe scissors
magazine pictures
markers
2 large index cards

Where: at a table then on the floor

How: With your children, go through magazines and look for pictures of items that are safe for children to play with (most toys, towels, plastic kitchen items) and items that are dangerous to play with (cleansing products, make-up, sharp or breakable items). Cut these pictures out of the magazines and put them in a pile on the table.

On one index card, draw a smiling face; on the other, draw a sad face. Move all of the pictures, the index cards, and the children to the floor. Put the cards face up on the floor. Spread out the pictures so the children can see all of them. Have the children take turns choosing a picture and telling whether the item is safe or dangerous. Have each child place the picture in a line below the appropriate index card.

Variation:
Talk about items in your surroundings. Discuss which are safe and which are not.

This activity also helps children learn about:
keeping themselves safe.

Sorting Circles
A Quick Trick with Juice Can Lids and Wrapping Paper

Gather These Materials:
lids from frozen juice cans
pencil
large index card
child-safe scissors
several patterns of wrapping paper
glue
magnetic tape (available at craft or hardware stores)

Where: on a steel (magnetic) cookie sheet, metal file cabinet, refrigerator door, or other magnetic surface

How: To prepare the activity: Gather a number of frozen juice lids. Trace around one lid onto an index card. Remove the lid. Now draw a line 1/8 inch (.3 cm) inside the tracing. Cut on the inner line, making a cardboard circle slightly smaller than the juice can lid. This is your template. Use the template to trace several circles on each kind of wrapping paper. Have your children help you cut out the circles, then glue one circle on the top of each lid. Let dry. Attach a strip of magnetic tape on the back of each lid.

To play: Give a child the decorated lids and have her place them on a magnetic surface. She will enjoy manipulating them and may form them into shapes, snaky lines, or other configurations. Suggest that she find the ones that are the same and put them together. Have her tell you how she chose each group of lids.

Variation:
Collect 25 lids and divide them into five sets. Decorate each set of five with a different kind of paper. Have the children arrange the lids in rows from top to bottom so that each column has matching lids. Help them see the pattern they have created with the five rows. Reading from left to right, they have five repeating rows of pattern.

This activity also helps children learn about:
creating patterns.

Quick Tricks for Math ©2001 Monday Morning Books, Inc.

Sorting My Cereal
A Quick Trick with Cereal

Gather These Materials:
waxed paper, 1 sheet for each child
cereal in different colors
small plastic bowls, 1 for each child (recycled margarine tubs or yogurt containers work well)
milk in small plastic pitchers
plastic spoons, 1 for each child

Where: at a table

How: Have your children wash their hands before beginning this activity. Put a piece of waxed paper the size of a place mat on the table in front of each child. Pour some cereal into each small bowl and give a bowl to each child. Encourage the children to find all the cereal pieces of one color and put them together on the waxed paper. Have them continue to group the cereals by color. After they have sorted all of their cereal, let them put the pieces back in their bowl, add a little milk, and eat them.

Variation:
Use cereal of different shapes and have the children sort by shape.

This activity also helps children learn about:
comparing, if they count each group to see which color has the most and which has the least.

Dear Parents,

We have been practicing the math concepts of sorting and classifying items. You can do this at home, too. When your child helps you fold laundry, she can place her clothes in one place and yours in another. She can put her socks and underwear into the correct drawers.

The skills of sorting and classifying help children develop logical thinking and stronger powers of observation.

Take advantage of the sorting and classifying opportunities that are all around you. Let your child help you load the dishwasher with unbreakable items (tall things go here, shorter items go here), put away unbreakable kitchen items (pots go here and dish towels go there), and find items in the grocery store. The "Quick Trick" below is one that your child can repeat again and again, improving with each experience.

Quick Trick with Flatware

Gather These Materials:
various kinds of flatware (spoons, forks)
flatware tray

Where: in the kitchen

How: Have your child wash her hands thoroughly, then let her help you put away flatware. Give her the flatware tray and the flatware that needs to be put away. She can sort the items into their appropriate places. Teach your child this song to sing while she puts the flatware in its special places:

Tune: "The Farmer in the Dell"
I put the <u>forks</u> away,
I put the <u>forks</u> away,
I know where the <u>forks</u> go,
I put the <u>forks</u> away.

Encourage her to continue singing as she puts other items away. She can change the underlined words to make the song appropriate to her task.

SHADES

The best way for children to learn about shapes is by using them in art and cooking activities, sorting them, and manipulating them in other ways. Use the shapes' names just as you do the names of other items. Ask a child to hand you the oval cookie cutter or the circle cookie cutter while you are making modeling dough cookies together.

Talk with your children about the attributes of the shapes. Describe them as having a certain number of sides and corners—for example, triangle, three sides and three angles. Be certain the children are able to count before you describe the numbers of the shapes' sides and corners.

Making Shapes
A Quick Trick with Cotton-Tipped Swabs

Gather These Materials:
cotton-tipped swabs, 12 for each child

Where: at a table

How: Give each child 12 swabs. Have the children arrange four of these into a square shape. Discuss the four sides and four corners, and how all four sides are the same size. Now have them use eight swabs to make a square shape beside the first square. Again mention the four equal sides and the four corners.

Have each child use three swabs to make a triangle. Discuss the three sides and three angles. Then have each child use four swabs to make a triangle beside the first. How are these triangles the same and how are they different?

Variation:
Add modeling dough to this activity and challenge the children to make three-dimensional figures by pinching balls of dough around both ends of the swabs and building with the swabs in many directions. Add more swabs by sticking them into dough balls already in place on previous swabs. When the children finish building, have them look for geometric shapes.

This activity also helps children learn about:
how parts go together to create a whole.

My Square Book
A Quick Trick with Square Paper

Gather These Materials:
adult scissors, for adult use only
plain paper cut into 4-inch (10-cm) squares, 8 sheets for each child
markers or crayons in 8 different colors
4-inch (10-cm) squares of construction paper, 2 sheets
for each child
stapler and staples
fine-line permanent marker

Where: at a table

How: Do this phase of the activity in two parts.
   Part 1: Give each child four squares of plain paper and four markers/crayons of different colors. Have the children color each paper a different color.
   Part 2: Give each child four squares of plain paper and four markers/crayons of colors he has not yet used. Have the children color each paper a different color.
   Help each child gather and stack his colored pages. Have him put one piece of construction paper on the bottom of the stack and one piece on the top. Have him construct a book by stapling the stack of papers along one edge. Use the marker to write the color and shape name on each page of each book. Write "My Square Book" on each book cover. Encourage the children to "read" their books to each other.

Variation:
Make similar books for other shapes: circle, triangle, hexagon, trapezoid, and so on.

This activity also helps children learn about:
how books are made.

Clowning Around
A Quick Trick with a Paper Plate

Gather These Materials:
solid-color paper plate
adult scissors, for adult use only
construction paper
glue or glue stick
crayons or markers (optional)

Where: at a table

How: The adult cuts small shapes (circles, half circles, diamonds, rectangles, triangles, squares) from the construction paper. Give each child a paper plate to use as a clown's face. Have each child choose from the shapes and glue them on as features for the face. Provide large construction paper triangles for each clown's hat. Have each child decorate his hat with some remaining shapes or with crayons or markers before gluing it on his clown's head.

Variation:
If your children are able to cut with success, let them cut out the shapes for the clowns' faces. If desired, provide shape stickers such as dots and stars for decorating the clown's hat.

This activity also helps children learn about:
how shapes go together to make a different image.

Triangles and Rectangles
A Quick Trick with Paper

Gather These Materials:
4-inch by 2-inch (10-cm by 5-cm) rectangle of blue felt
4-inch by 2-inch (10-cm by 5-cm) rectangle of yellow felt
flannel board
marker
ruler
adult scissors, for adult use only
4-inch by 2-inch (10-cm by 5-cm) blue construction paper
rectangles, 1 for each child
4-inch by 2-inch (10-cm by 5-cm) yellow construction paper
rectangles, 1 for each child
scissors, 1 pair for each child
glue or glue sticks

Where: in front of a flannel board to begin, then at a table

How: In small groups, show children how the felt rectangles are identical in size. Place them side-by-side on a flannel board. With a marker, draw lines dividing the yellow rectangle, diagonally, into four triangles. Cut on the lines and show the resulting triangles to the children. Ask a child to place the yellow triangles on top of the blue rectangle to re-form a yellow rectangle. Then give each child a turn re-forming the triangles.

   At a table, give each child a paper rectangle of each color. Have them fold their yellow rectangles diagonally in one direction and crimp the paper along the fold. Have them open their papers, then fold them diagonally in the other direction and press along the fold. Have children open the rectangles and cut along the folds to create four yellow triangles. Have children place the yellow triangles on top of the blue rectangle, re-forming a yellow rectangle. Let them glue the yellow triangles in place, forming a rectangle that is blue on one side and yellow on the other.

Variation:
Repeat the activity using two colors of 4-inch (10-cm) squares. Cut one square diagonally in both directions to see how four identical triangles can form a square.

This activity also helps children learn about:
how parts go together to become a whole.

Make a Picture
A Quick Trick with Geometric Paper Shapes

Gather These Materials:
paper geometric shapes: circles, rectangles, squares, triangles of various sizes (cut your own or purchase them already cut)
plain paper
glue or glue sticks

Where: at a table

How: Provide your children with precut geometric shapes. Encourage them to use these to form other two-dimensional things. They may make cars, trees, houses, animals, people, TV sets, or even a whole neighborhood. Let them glue these items onto the plain paper, if they desire.

Variation:
Add to the shapes some trapezoids, hexagons, ovals, and diamonds.

This activity also helps children learn about:
how parts go together to become a whole.

Shapely Snacks
A Quick Trick with Food

Gather These Materials:
various foods in different shapes (see below)

Where: any place you eat snacks or meals

How: Plan a snack or a light meal around foods of one shape. Let the children help prepare and serve the foods. Discuss the shapes while the children are preparing and eating the foods. Here are some suggestions:

　　Circles: round crackers, sandwich cut with a round cookie cutter, banana slices, muffins, cookies, cucumber slices, thin carrot slices, tomato slices, pickle slices, hamburger.

　　Triangles: triangular crackers, triangular chips, sandwich and cheese slices cut into triangles, brownies or bar cookies cut into triangles.

　　Squares: brownies, sheet cake or bar cookies cut into squares, square crackers and graham crackers, cheese cut into squares, luncheon meat cut into squares, meat loaf.

　　Rectangles: rectangular crackers and granola bars, whole graham cracker, cookie/brownie rectangles, sandwich cut into rectangles, meat loaf.

Variation:
For a sandwich of circles and squares, have the children use a butter knife to spread peanut butter on a square slice of bread. Let them slice half a banana into circular slices. Have them place the banana slices on the peanut butter and top all with another square slice of bread. Now they can fill up on circles and squares. Caution: Do not use peanut butter if any of the children are allergic to peanuts!

This activity also helps children learn about:
eating a variety of foods.

Hmmm!

Matching Halves
A Quick Trick with Felt Shapes

Gather These Materials:
pencil
adult scissors, for adult use only
felt in 1 color only
plastic zipper bag

Where: at a table to prepare the activity
anywhere to play the activity

How: To prepare the activity, draw and cut out one each of various shapes: square, triangle, rectangle, circle, oval, hexagon, diamond. On each shape, draw a line dividing the shape in half so that both halves are identical. Cut along the lines. Store the felt game pieces in a plastic zipper bag.

   To play the game, a child spreads out the contents of the bag. She finds matching halves of a shape and places them so that they form a completed shape. Encourage her to name all the shapes she makes.

Variation:
For younger children, use a different color for each whole shape.

This activity also helps children learn about:
paying attention to detail.

Dear Parents,

We have been doing activities at school to learn about different shapes. Your child has learned ways to cut shapes to form other shapes. When you make sandwiches together, he can show you what he has learned. The "Quick Trick" below will guide you.

A Quick Trick with a Sandwich

Gather These Materials:
soft sandwich spread (egg salad, peanut butter, pimento cheese spread, or tuna salad)
square-shaped bread, 2 slices for each sandwich
plastic knives or butter knives

Where: at a table

How: Invite your child to help you prepare simple sandwiches for a family meal. Use square-shaped bread and talk about the shape. Work with your child as he spreads a soft filling on half of the bread slices. Top the spread halves with the remaining plain bread.

Ask your child to show you how to cut one of the square sandwiches in half diagonally to form two triangles. Ask if he can show you how to cut another square sandwich into smaller squares. Let him cut any remaining sandwiches as he desires, and have him tell you what shapes he has created.

PATTERNS

Patterns occur all around us. Summer becomes autumn which becomes winter and then spring, then the pattern repeats again and again.

Day becomes night, which is followed again by day and night. Days of the week follow the same pattern week after week and the months follow each other throughout our lives.

In math, if we write out the "fives" multiplication table, we see a pattern: the total increases by five each time we multiply the number 5 by the next numeral in the numeric sequence. There are similar patterns in all of the multiplication tables as well as in addition, subtraction, and division.

Through playing with patterns, children become aware of the patterns around them. Playing with patterns helps children learn to use existing information to predict what will happen next. Recognizing and understanding patterns help people understand math's predictable qualities.

I Spy a Pattern
A Quick Trick with Clothing

Gather These Materials:
child wearing a garment with a pattern of regular stripes

Where: anywhere

How: When you see a child in the group wearing a two-tone striped shirt, say, "I Spy a Pattern." Have the child stand in front of the group while you point out the shirt's pattern: blue/white, blue/white. Have the group look for patterns on each others' clothing. Remember that patterns can run horizontally or vertically.

Variation:
Look for patterns on wallpaper, wrapping paper, table cloths, and other environmental surroundings.

This activity also helps children learn about:
using their powers of observation.

Moving Patterns
A Quick Trick with Movement Activities

Gather These Materials:
a group of children
bell

Where: anywhere

How: Have children gather in a group. Chose two to come to the front of the group. Each child selects a different way to move her body. (Suggestions include galloping, marching, hopping or standing in place while swinging an arm, lifting shoulders, and nodding.) The two selected children show their motions to the group.

Now designate one of these two children to lead the group while they copy her motion for 10 seconds. At a signal from you (the ringing bell), the second child leads the group while they copy her motion for 10 seconds. Ring the bell again, and the first child leads again. Continue the movements in a pattern for about a minute. Choose new leaders and begin again.

Variation:
Repeat the activity while playing music. Use different kinds and tempos of music each time you repeat the activity.

This activity also helps children learn about:
listening.

Quick Tricks for Math ©2001 Monday Morning Books, Inc.

Button, Button
A Quick Trick with Buttons

Gather These Materials:
a variety of buttons

Where: at a table or on the floor

How: In a small group, give the children a variety of buttons and challenge each child to use some of them to form a pattern. Have each child explain her pattern to the group. At first, children usually arrange a pattern based on button size or color. As you repeat this activity over time, and patterning skills increase, some may focus on the number of holes in the buttons or arrange a pattern of fancy and plain buttons.

Variation:
For a lasting pattern, have each child glue her button pattern onto a strip of cardboard. Write her words on the strip as she "reads" the pattern.

This activity also helps children learn about:
sorting and classifying as they choose the buttons for their patterns.

Days of Our Lives
A Quick Trick with Index Cards and Magnets

Gather These Materials:
7 large index cards
marker
magnetic tape

Where: a magnetic surface

How: Write one day of the week on each of seven index cards. Put two strips of magnetic tape on the back of each card. Have seven children, in turn, place the cards on the magnetic surface. Have them arrange the cards in a column, from top to bottom, with today's day at the column's top. Do this by giving the first child the card with today's name on it and giving each succeeding child the next days' names, in order. Name the days as they are displayed, and then have a child point to each card in turn while together the group says the days' names.

Each day at large group time, supervise as a child removes the top card (yesterday) and places it at the bottom of the column. Then, have her point to each card in order while the group says the days' names together. Remind children that the top word is today's name.

When a day (or days) have been missed due to weekends or holidays, name and read the day's name while a child moves it to the bottom of the column.

Variation:
Over time, as children learn the repeating pattern, talk about yesterday and predict what day comes tomorrow. Talk, too about the "day before yesterday" and the "day after tomorrow."

This activity also helps children learn about:
the passage of time.

Quick Tricks for Math ©2001 Monday Morning Books, Inc.

Pretty Patterns
A Quick Trick with Classroom Materials

Gather These Materials:
small quantities of a variety of classroom items (see below)
baskets or shopping bags, 1 for each child

Where: on the floor

How: In each basket, place approximately 20 classroom items of the same kind. In one, you might place 20 pom-pom balls or 20 snap-together cubes in several colors or sizes. Other suggested items include colored, vinyl-clad paper clips; discarded keys; buttons; caps from discarded markers; bottle caps; crayons; variety of dried beans.

Give a basket of items to each child. Challenge the children to use their items to make a pattern. When they have succeeded, have each show her pattern to a friend and tell how she made it. Then the friend shows and tells about his pattern. Have the children put their items back in their baskets then swap baskets with a friend. Challenge them to begin again, and repeat the activity until attention lags. Some children will want to continue using the items and making patterns long after others have moved on to other activities. If possible, allow them to do so.

Variation:
Group the children in pairs. Have one child of each pair begin a pattern with her items. Now challenge the other child to continue the pattern.

This activity also helps your child learn about:
cooperating and sharing with a friend.

Party Pattern
A Quick Trick with Construction Paper

Gather These Materials:
ruler
pencil
blue construction paper, 1 sheet for each child
yellow construction paper, 1 sheet for each child
child-safe scissors
tape or a miniature stapler for each child

Where: at a table

How: Do the following with each child: With a ruler, draw lines across the construction paper, lengthwise, about 1 inch (2.5 cm) apart. (If desired, substitute other colors of construction paper.) Give a sheet of both colors to a child. Have her cut on the lines to form 1-inch-wide (2.5-cm) strips of paper. Have her lay the strips of paper in a simple pattern (blue, yellow, blue, yellow).

Show the child how to form a circle from the first strip in the row. Have her tape or staple the circle closed. Now have her put the next strip through the circle and bend that second strip into a closed circle. Have her tape or staple the second strip closed. Let her continue working in that manner until she has a long length of colorful party chain. Repeat this activity with every child. Use the children's chains to decorate for a simple party. Learning about patterns is a good reason for a celebration!

Variation:
Repeat the activity on another day using two colors, but form a different repeating pattern: blue, yellow, yellow, blue, yellow, yellow; or blue, blue, yellow, blue, blue, yellow.

This activity also helps children learn about:
hand-eye coordination.

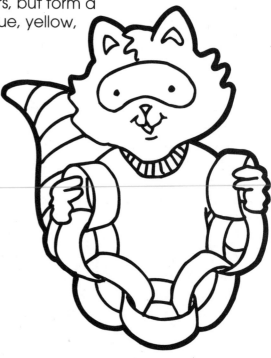

Quick Tricks for Math ©2001 Monday Morning Books, Inc.

Dear Parents,

If your child has been pointing out patterns all around you, you've probably realized that we are exploring patterns at school. This is a part of learning about math and science.

Many mathematical procedures are made up of patterns. For example, when we count by fives, we repeat the pattern of "adding five" to each consecutive step.

The more we learn about patterns, the more predictable math becomes. People who understand patterns become strong mathematicians.

By playing with patterns, creating them, and discovering them, children become aware of patterns all around them. Playing with patterns helps children learn to use existing information to predict what will happen next. Recognizing and understanding patterns help people understand math's predictable qualities.

Take advantage of your child's interest in small, sticky-note pads and let her practice making patterns with the "Quick Trick" below.

A Quick Trick with Sticky Notes

Gather These Materials:
same-size sticky notes in various colors

Where: at a table

How: Show your child how to peel and stick "sticky notes" of two different colors onto the table top to make a pattern (for example, blue, yellow, blue, yellow, blue, yellow). Give her two colors of sticky notepads and challenge her to use them to form a pattern. Repeat the activity with pads of two other colors. Now challenge her to make a different pattern using two colors (for example, blue, yellow, yellow, blue, yellow, yellow, blue, yellow, yellow).

Let your child repeat this activity as often as she desires. She'll be learning more and more about patterns each time. She can reuse the same sticky notes again and again. If the sticky substance wears out, provide a glue stick and let your child glue the little notes in a pattern on a sheet of paper.

Young children have a fuzzy concept of the passage of time and the use of the words *yesterday, tomorrow,* and *today*. With our help, they can become more aware of time's passage.

A child must be able to read numerals before he can learn to "tell time." Earlier Quick Tricks helped children learn to read numerals; tricks in this section focus on the concept of time's passage.

Help children become aware of clocks and calendars and the fact that they mark the passage of time. Place one or more battery-operated clocks where the children can see them and, if desired, reset their time. Have another large clock hanging above their reach where they can see it and you can refer to it as needed. Hang a class calendar where all can see it. Record special events on it each day.

Calendar Days
A Quick Trick with a Calendar

Gather These Materials:
a current calendar
a crayon

Where: where all of the children can see the calendar

How: Place the calendar where everyone can see it. Every day, have a different child mark off the current day. Say the name and date of the current day as the child marks it. Explain to the group that when they arrive the following morning, it will be the next day, and say that day's name and date. From time to time, "read" or count all of the days that have passed in the current month.

Variation:
When a special occasion is approaching, place a picture on the proper calendar day to designate the event. Each day have the children count how many days until the event.

This activity also helps children learn about:
counting.

Disappearing Beans
A Quick Trick with Beans

Gather These Materials:
large dried beans
small plastic jar

Where: on a table where all of the children can see

How: To help children count down the days until a special occasion (a field trip, a party, a special visitor), count with them the number of calendar days until the event. Together, count out that many large, dried beans. Have a child put these in a plastic jar and place it where all can see it. At the same time each day (large group time), have a different child remove one bean from the jar. The children can watch the diminishing number of beans as the special day gets closer. When they remove the final bean, they'll know that the special day is here.

Variation:
If desired, count the remaining beans each day and then return them to the jar. Have the children predict the number of remaining beans.

This activity also helps children learn about:
subtracting by one, when they count the remaining beans each day.

Set the Timer
A Quick Trick with a Kitchen Timer

Gather These Materials:
kitchen timer

Where: various places

How: Show the children a kitchen timer and let them play with it, wind it, and hear the bell ring. Show them how to set the timer to a numeral. Explain that in that number of minutes, the timer will ring.

At the beginning of large group time, set the timer for the amount of time you have planned for large group activities. When the timer sounds, large group time is over. This will help children understand that a timer measures an amount of time and then announces when that designated amount of time has passed. Set the timer in a similar way for Outdoor Time. When the timer sounds, it's time to go inside.

Variation:
Tell your children when it's five minutes before clean-up time. Let a child set the timer for five minutes. Leave the timer in view so all children can observe its changes. When the timer sounds, they'll know to begin putting their toys away.

This activity also helps children learn about:
paying attention to auditory signals.

Large group time is over!

Sinking Times
A Quick Trick with Jar Lids

Gather These Materials:
large nail
hammer
metal or plastic jar lids of similar sizes
water table or dishpans containing water

Where: at a table (for preparing materials)
at the water table (for using the materials)

How: The adult does the following: Using the nail and hammer, punch one hole in several jar lids, two holes in others, and three holes in others. Place these items in the water table.

Show the lids to the children. Encourage children to observe what happens to the lids when they attempt to float them on the top of the water. Which takes longer to sink? Which takes less time? Can they explain why this happens?

Help children count with a steady beat how long it takes each lid to sink. Offer more jar lids of various sizes to add opportunities for learning. Have them predict which lids will take more or less time to sink.

Supervise carefully when children are playing with water.

Variation:
In warm weather, children can also enjoy this activity outdoors in a wading pool.

This activity also helps children learn about:
sinking and floating.

Watch the Clock
A Quick Trick with a Clock

Gather These Materials:
working clock (not digital)
colored tape

Where: anywhere the children can easily see the clock

How: After you have designated a time for outdoor play, place two pieces of tape on the clock to mark where the clock's hands will be at the appointed time. Mark the "long" hand place with a long, skinny piece of tape and the "short" hand with a small circle you have cut from the tape. Call your children's attention to the clock as outdoor play time approaches, and have them look at the clock when the appointed time arrives.

Variation:
With a marker, draw a clock face on a paper plate. Use a marker of one color for the minute hand and a different color for the hour hand. Create a clock set for the start of outdoor play time. Show this clock face to the children and discuss what it shows. Hang this paper plate clock beside the classroom clock and remind children to look at the clock from time to time to see if the classroom clock's hands indicate it is time for outdoor play. Using paper plates of other colors, make clock faces that show the times of other important activities: lunch time, dismissal time, and so on.

This activity also helps children learn about:
noticing changes.

ESTIMATION

Children enjoy opportunities to build their estimation skills. For them it's a game. Provide opportunities for the children to estimate length, width, weight, height, amount, and more. These experiences will strengthen their skills.

Clips and More Clips
A Quick Trick with Paper Clips

Gather These Materials:
jumbo, vinyl-clad paper clips, 20 for each child and 20 for yourself

Where: anywhere

How: Show the children how to slide one paper clip into another to join them together. Show them "chains" of eight connected clips, six connected clips, four connected clips, and two connected clips.

Select a child to stand at the front of the group. Ask the children how many clips they think it will take to make a chain that reaches from the child's fingertips to his elbow. Give paper clips to the children and have each make a chain that he thinks is the right length. Let several children measure their chains against the child's outstretched arm. If one chain is the right size, count the number of clips in that chain. Have the children count the clips in their chains. Ask how many children had the same length chain. How many had more clips? How many had fewer? They can also determine this by measuring their chains beside the chain that is the right length.

Use this same technique to estimate other amounts.

Variation:
At another time, repeat the activity using smaller paper clips. Have each child make two chains to estimate the same distance—one with the larger clips and one with the smaller clips. Over time children will begin to realize that it takes more of the smaller-size clips to make a chain equal in length to one made of larger but fewer clips.

This activity also helps children learn about:
measurement.

How Big?
A Quick Trick with a Watermelon

Gather These Materials:
watermelon (or other large fruit)
several balls of yarn
child-safe scissors, several pairs
marker
masking tape

Where: at a table or on the floor

How: Show the children a watermelon or other large fruit (canta-loupe, honeydew melon). Make the yarn and scissors available to the children. Have each child cut a piece of yarn the length he thinks is the distance around the watermelon at its center. Write each estimator's name on a piece of tape and attach the tape to his yarn. After everyone's yarn is labeled, let the children, in turn, wrap the yarn around the melon. Whose yarn is closest to the watermelon's girth?

Variation:
Use yarn to estimate the circumference of other items like an orange, a ball, a pumpkin, or a tree trunk.

This activity also helps children learn about:
measurement.

How Tall?
A Quick Trick with Yarn

Gather These Materials:
full-length mirror
yarn
scissors

Where: anywhere

How: Have a child look into a full-length mirror to study his height.
Now give him a ball of yarn and have him pull out the amount of
yarn that he thinks will equal his height. Help him cut the yarn at
that point. In front of the mirror, hold one end of the yarn at the top
of his head and gently straighten the other end while placing it
toward his feet. (Avoid pulling on the yarn because that will stretch
it.) Ask him if the yarn is the same length as he is tall, or if it is shorter
or taller than he is. If the yarn was not the correct length, help him
measure and cut off a piece of yarn the same length as he is tall.
Encourage him to straighten it out on the floor to see its length.
Repeat with the remaining children.

Variation:
Have the children compare their yarn lengths with those of other
children. Encourage them to stand in front of the mirror with a friend
to see who is taller or shorter.

This activity also helps children learn about:
comparison, as a child compares his yarn's length
with his own height.

Inside and Outside
A Quick Trick with Kitchen Containers

Gather These Materials:
a variety of plastic kitchen containers in many sizes
a variety of empty food boxes (cereal, gelatin, pasta) in many sizes

Where: on the floor

How: Gather containers in a variety of sizes and show them to the children. Ask one child to choose a container, then challenge him to guess (estimate) which other containers or boxes will fit inside it. Have him check his predictions. Now have him choose one other container that he thinks his container will fit in and have him check this prediction. ·

Continue with other children. Ask the group if they agree with the child's predictions. After the activity, place all of the containers and boxes in the Dramatic Play Area so that the children can continue to explore them.

Variation:
Challenge a child to choose a container that his item will NOT fit in.

This activity also helps children learn about:
hand-eye coordination.

VOLUME

Volume is a measure of the amount of a solid, liquid, or gas. Long before young children can understand pints, quarts, or liters, they will begin to understand that some containers hold more sand than others or that one of this container holds as much water as two of these containers.

Provide your children with many activities to experience more, less, and the same volume. They can explore the volume of sand, water, and salt when the items are in plastic containers or heavy cardboard boxes. A water table, dishpan, and wading pool are good places to explore water's volume. Furnish plenty of opportunities for hands-on activities, and describe things using words like *more, less,* and *the same as.*

How Many?
A Quick Trick with Styrofoam Chips and Boxes

Gather These Materials:
dishpan or other, similar-size plastic container
Styrofoam chips or Styrofoam "peanuts"
small boxes (from gelatin/pudding mixes, jeweler's boxes, toothpaste cartons, and so on)
small plastic containers (jars or deli containers)
paper and pencil

Where: on the floor

How: Working one-to-one with a child, give him a large box of Styrofoam chips and a number of small boxes and plastic containers of various sizes. Have him predict how many Styrofoam pieces the smallest container will hold and have him record the predicted number. Now have him fill the container and then spill out and count those Styrofoam pieces. Continue in the same manner with the remaining boxes.

Variation:
Have the child spill out the contents of two different boxes and compare the pile of chips made by each.

This activity also helps children learn about:
estimating.

How Much Dough?
A Quick Trick with Modeling Dough

Gather These Materials:
measuring cups
commercially available modeling dough

Where: at a table

How: Have a child choose two or three plastic measuring cups that are noticeably different sizes. Have him fill each cup to the top with modeling dough. Talk about which cup has more dough, the most dough, less dough, and the least dough.

Encourage him to empty a cup and make something with that dough. Now have him empty another and make something with that amount, then have him repeat the activity with the third cup. Talk about the sizes of the different items he made. Talk about other things he'd like to make. How much dough does he think he'll need to make each item? Let him measure out an amount of dough and see if it is the amount he needs. Repeat the activity with other children.

Variation:
Have the child break off two lumps of modeling dough. Ask him which lump has more/less dough in it. Have him put each lump in a measuring cup of the nearest size and see which cup is larger/smaller. Don't try to explain the concept of more/less volume to young children. It takes many repetitions of this and similar volume activities before they will begin to develop an understanding of which amount is more and which is less. Just let them have fun manipulating the material. After many repetitions over time, the understanding will happen.

This activity also helps children learn about:
math vocabulary: more, less, most, least.

Quick Tricks for Math ©2001 Monday Morning Books, Inc.

Rice Is Nice
A Quick Trick with Rice

Gather These Materials:
plastic cloth (shower curtain/table cloth)
5-pound (2.25-kg) bag of rice
plastic storage bin with tight-fitting lid
plastic measuring cups of various sizes
spoons and scoops of different sizes
plastic jars, bowls, and containers of various sizes

Where: on the floor or at a table

How: Spread the plastic cloth on the floor. Pour rice into the bin until it is about 4 inches (10 cm) deep. Place the measuring cups and spoons in the bin. Give the bin to the children to explore freely. Remind them that all rice must stay in the bin, the containers, or on the plastic cloth. As they fill the containers with rice and pour it from one container to another they'll experience that different size containers hold different amounts of rice. They may discover that it always takes two 1/2-cup (.12-l) measuring cups to fill one 1-cup (.25-l) measuring cup and four 1-cup (.25-l) measures to fill a 1-quart (1-l) bowl. Discuss this with them occasionally, but don't turn the fun into "a lesson."

For a quick clean up, move the bin off the plastic cloth. Pick up the cloth by its four corners and pour any spilled rice back into the bin.

Variation:
When children's interest lags, add everyday items (bottle caps, jar lids, small toys, large buttons, golf balls) to the bin, hiding most of them. The children will enjoy finding them and hiding them again.

This activity also helps children learn about:
focusing attention.

Turn Up the Volume
A Quick Trick with Popcorn

Gather These Materials:
unpopped popcorn (not microwave popcorn)
see-through measuring cup
popcorn popper (a hot-air popper works best)

Where: at a table or on the floor—where everyone can see

How: Following the directions on your favorite brand of popcorn, measure out the proper amount of unpopped kernels. Show the children the unpopped kernels in the measuring cup. Give a child a crayon and have him mark the kernels' level in the cup. Now pop the corn according to the manufacturer's directions. Pour the popped corn into the measuring cup up to the previous level of unpopped corn. Discuss the difference in the unpopped corn and the popped corn. Measure the remaining popped corn using the original measuring cup. As a group, count how many times you can fill the cup with popped corn. Discuss with the children that heat changed the corn to a form that takes up more volume.

Variation:
If you can only use microwave popping corn in prepackaged bags, show the children two bags of unpopped corn. Let them determine that the bags are similar in volume. Pop one bag of corn and then compare the unopened popped bag with the unpopped one. Discuss the change in volume.

This activity also helps children learn about:
measurement.

CONSERVATION OF NUMBER

A child demonstrates that he has the ability to "conserve" number when he can see the same number of items in different configurations and recognize that their number is constant. For example, a child sees and counts five buttons placed in a straight line on a piece of construction paper. When he or an adult moves the buttons around the paper and arranges them in random fashion, the child who conserves number will say there are still five buttons. The child who does not yet conserve number needs to count each time the buttons are rearranged.

Changes and More Changes
Quick Trick with Building Blocks

Gather These Materials:
building blocks or snap-together blocks

Where: on the floor

How: Select 10 to 12 blocks for each child. Give each child in a small group pieces of the building material that are <u>identical</u> to everyone else's materials. Have everyone count the blocks together. Do they agree that each has the same number and identical blocks?

Challenge everyone to build something with their blocks. Have each child show and tell about what he has built. Do they still think that everyone has the same blocks?

Now have everyone take apart the constructions and build something different. Again, have them show and tell about their constructions and discuss whether everyone has identical blocks and identical numbers of blocks.

Variation:
Give each child six different paper shapes. Each child's group of shapes should be identical to every other child's group of shapes. Provide glue sticks and plain paper. Challenge each child to glue his shapes to the paper to make a picture. Display the pictures and discuss with the children the fact that each different picture has the same number and type of shapes.

This activity also helps children learn about:
counting and comparing.

Fix the Sticks
A Quick Trick with Craft Sticks

Gather These Materials:
craft sticks, 4 for each child

Where: on the floor

How: Give each child four craft sticks. Have each arrange his sticks to make a square. Challenge him to move the sticks so they form a diamond. Have him use all four sticks to make the letter C. Repeat with I, J, K, L, M, T, W, X, and Z using all four sticks for each letter. Talk about how many sticks the child uses each time and the fact that although the designs change, the number of sticks remains the same.

Repeat the activity with other numbers of sticks and other configurations. Say this rhyme with the children each time they move their sticks to a new configuration:

> I like to play with craft sticks,
> It's such a dandy game.
> No matter how I change the shape,
> The number stays the same.

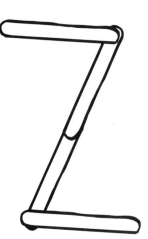

Variation:
Give each child six sticks and let each decide what to make. He may decide to make a star, house, flower, rectangle, or other items with the six sticks. After each creation, have each child count his sticks and discuss the fact that although the designs change, the number of sticks remains the same.

This activity also helps children learn about:
letter formation.

Quick Tricks for Math ©2001 Monday Morning Books, Inc.

Ghosts in the Meadows
A Quick Trick with Dried Beans

Gather These Materials:
fine-point black, permanent markers, 1 for each child
large, dried lima beans
green construction paper, 2 sheets for each child

Where: at a table

How: To make the game pieces, closely supervise while the children make two "eyes" on each bean so the beans resemble ghosts. Give each child two same-size sheets of construction paper and have him put five beans on each sheet. Have him put the ghosts close together on one sheet and far apart on the other. Discuss the ghosts in the meadows. Do both meadows have the same number of ghosts? (Many children, even after counting, think that the meadow with the spread-out ghosts has more because it looks like more. Don't worry about this; just keep playing the game.)

Have the children empty their meadows and begin again. Each time, have them put the same number of ghosts in each meadow and talk about how many ghosts are in each meadow.

At another time, have the children put ghosts in one meadow and count them. Now have them move those same ghosts to the other meadow. How many ghosts do they think are in this meadow? Have them count the ghosts they see. Repeat this activity several times.

Variation:
Draw smiling faces on bottle caps and use them instead of ghosts. They can be happy children at two playgrounds.

This activity also helps children learn about:
counting.

ONE-TO-ONE CORRESPONDENCE

When a child selects two shoes from the dress-up clothes that he wants to wear, he demonstrates his knowledge of one-to-one correspondence. By selecting two shoes for two feet, he is pairing items in a one-to-one relationship.

It takes a lot of experience with hands-on tasks before a child can translate this awareness from his own body (two feet, two shoes) to items outside himself. Over time, setting the table (one napkin to each person), playing with blocks (one roof for each house), he internalizes the idea of "one of these" for each "one of those." Children must understand one-to-one correspondence before they can determine whether two groups of items have the same number.

Edible Zoo
A Quick Trick with Animal Crackers

Gather These Materials:
paper plates, 1 for each child
graham crackers, 1 for each child
peanut butter
butter knives, 1 for each child
animal crackers, 4 for each child

Where: at a table

How: Have the children wash their hands. For each child, divide a graham cracker into four small rectangles, and have the child use the butter knife to spread peanut butter on each piece of cracker. These will be the animals' areas in the zoo. Have children choose one animal cracker to stand on each graham cracker "area." If the crackers won't stand up, lay them flat. Encourage the children to arrange the crackers into a zoo on their plates. When they are through playing, they may eat their zoos as a snack. <u>Caution:</u> Do not use peanut butter if any child in your group has a peanut allergy.

Variation:
In case of peanut allergies, use applesauce for all children. Lay the animals flat in the applesauce.

This activity also helps children learn about:
small motor coordination.

Serving Bowls
A Quick Trick with Modeling Dough and Pasta

Gather These Materials:
commercially available modeling dough
small uncooked pasta shapes, like seashells

Where: at a table

How: Show the children how to make walnut-size balls of dough.
Have each child push his thumb into a ball and pinch the dough
between his thumb and fingers. As he turns the ball in his hand and
continues pinching, he will form a small bowl.

   After he makes a number of bowls, have him count the bowls
and count out a corresponding number of pasta pieces. Encourage
him to put the pasta in the bowls to see if each bowl receives an
identical amount of pasta.

Variation:
At another time, put out other items for children to use while
creating with the modeling dough. Encourage all efforts.

This activity also helps children learn about:
small motor coordination.

SERIATION

Seriation means placing things in order using one criterion or several criteria—for example, having children line up according to height. It is a form of comparison, and a child must be able to see similarities and differences in order to seriate. Be certain the children have many experiences with comparisons before focusing on these Quick Tricks.

How I've Grown
A Quick Trick with Photographs

Gather These Materials:
4 or 5 pictures of yourself at different ages (for example, as a baby, toddler, schoolchild, and adult)

Where: anywhere you can comfortably sit with a small group of children and spread out the pictures in front of them

How: Show children the pictures of yourself taken at different ages. Show them in random order. Spread them face-up on the table. Ask a child to pick up the picture where you were youngest. Tell the group something about yourself when you were that age. Next, ask another child to find and pick up, among the remaining pictures, the picture where you were the youngest. Again tell something about yourself at that age. Continue in this manner until all of the pictures have been selected. Have the children arrange the pictures in order from youngest to oldest. Repeat the activity selecting and arranging the photos from oldest to youngest.

Variation:
Make photocopies of the pictures and mount the copies on cardboard. Laminate them for durability. Place the pictures where the children can arrange them in order as they act out "Teacher Growing Up" stories.

This activity also helps children learn about:
time concepts.

You were youngest in this one.

Stick Tricks
A Quick Trick with Sticks

Gather These Materials:
4 or more sticks of different lengths

Where: outdoors, on the ground

How: Help a child gather four or more sticks of different lengths. Lay them at random, side-by-side on the ground. Ask him to find the longest stick and place it in front of himself. Refer to the remaining sticks and ask him again to find the longest stick and place it to the right of the first. (Indicate the position with your hand.) Continue until he has placed the sticks in order from longest to shortest.

Mix up the sticks and repeat the activity arranging the sticks from shortest to longest. As the child gains skill, add more sticks. Repeat with other children.

Variation:
Gather sticks of different thickness. Have children arrange them in order from thinnest to thickest or from thickest to thinnest.

This activity also helps children learn about:
comparison.

Which Is Larger?
A Quick Trick with Kitchen Items

Gather These Materials:
a set of plastic nesting bowls

Where: on the floor

How: Show the children a set of plastic nesting bowls. Have a child arrange them in order from largest to smallest. Mix them up and have a different child arrange them from smallest to largest. Talk about the bowls using the words *small, smaller, smallest, large, larger,* and *largest*. Have another child nest the bowls inside one another. Do similar activities with nesting canisters, measuring cups, or cooking pots.

Variation:
Encourage the children to stack the bowls upside-down with the largest on the bottom and the smallest on top.

This activity also helps children learn about:
paying attention to details.

This is the largest bowl.

Quick Tricks for Math ©2001 Monday Morning Books, Inc.

Whose Shoe?
Quick Trick with Shoes

Gather These Materials:
shoes of noticeably different sizes

Where: on the floor where everyone can see

How: Place shoes of various sizes on the floor in front of the children. (Use single shoes, not pairs. If possible, have the shoe of a baby, a young child, an older child, and various-size adult shoes.) Have a child find the smallest shoe and place it in front of and to the left of the group. Point to the remaining shoes and ask another child to find the smallest remaining shoe and place it to the right of the first. Continue playing until the children have placed the shoes in order from smallest to largest. Have the children arrange the shoes from largest to smallest. Do the same activity with socks of various sizes.

Variation:
Place a matching number of various-sized shoes and socks in front of the children. Have them arrange each group in order from smallest to largest, matching the smallest sock with the smallest shoe, the next-smallest sock with the next-smallest shoe, and so on.

This activity also helps children learn about:
paying attention to details.

FRACTIONS

Young children are just beginning to build their concepts of fractions. When they divide one item into equal parts, or put those parts together to form the original item, they begin to see how equal parts fit together to make a whole. Another way of experiencing fractions is by dividing an amount (like six cookies) into equal amounts (for three people). Children easily understand when another person has more or fewer items than they do.

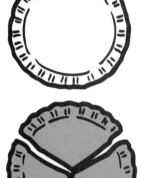

Circle Puzzles
A Quick Trick with Paper Plates

Gather These Materials:
paper plates, 4 for each child
markers in 4 colors, not black
ruler
black marker for teacher's use
blunt-edge scissors

Where: at a table to make the puzzles
at a table or on the floor to use the puzzles

How: Give a child four lightweight paper plates. Have him use markers to color each plate a different color. Using a black marker and a ruler, divide one plate into halves, one into thirds, and one into fourths. Do not mark the remaining plate. Have the child cut each plate along the dividing lines. Have him mix up all the pieces, then challenge him to put the plates back together. Talk about the numbers of sections on each plate and use the words *half, whole, third,* and *fourth.* Examine the puzzle pieces together. Challenge the child to tell you which has <u>larger</u> pieces: the puzzle with thirds or the one with fourths? Which has <u>more</u> pieces? Compare the puzzles with each other. Repeat with every child.

Variation:
Add one more colored plate to each child's collection. Draw lines on it dividing it into eighths and have the child cut on the lines. Challenge the child to find new ways to put plates together while mixing colors.

This activity also helps children learn about:
cutting skills.

Shapes Within Shapes
A Quick Trick with Paper Shapes

Gather These Materials:
pencil
ruler
red construction paper
scissors

Where: at a table or on the floor

How: To prepare the activity, draw a 4-inch (10-cm) by 6-inch (15-cm) rectangle on red construction paper. Draw a diagonal line from one corner to the other. Cut along the line, forming two triangles. Make a set for each child and one for yourself. Show the children how the triangles go together to form a rectangle. Take apart the demonstration rectangle and challenge a child to rearrange the triangles to re-form the rectangle. Now have all of the children arrange their two triangles into rectangles. The children begin to understand fractions as they see how the equal parts fit together to make the whole shape.

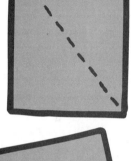

Variation:
For each child, prepare a 6-inch (15-cm) square on yellow construction paper. Draw two diagonal lines from corner to corner, forming an X in the center. Cut along the lines forming four triangles. Give each child a set of these pieces and keep a set for yourself. Show the children how the triangles go together to form a square. Challenge the children to rearrange their triangles to form squares.

This activity also helps children learn about:
shapes.

Note: This activity is similar to the Quick Trick "Triangles and Rectangles" on page 60. You can also use that trick to focus on fractions, and you can use this trick to focus on shapes.

The Money Game
A Quick Trick with Coins

Gather These Materials:
a plastic dish for each player
many coins

Where: at a table or on the floor

How: In a small group, give each player a dish with 15 coins in it. Put the remaining coins in the middle of the group where everyone can reach them. Show your children how to flip a coin and look at it to determine if it lands "heads up" or "tails up."

The first player flips a coin. If the coin is "heads up," he takes two coins from the center and puts them in his dish. If the coin lands "tails up," he must evenly divide the contents of his dish with all of the other players. (Show him how to do this by giving "one to this player, one to that player, and one to me.") Players can vote about what to do if he has fewer coins than the number of players. They may decide that he does nothing or that he must put all of his remaining coins back in the center. They may determine some other solution. The decision holds for the entire game. The winner is the player who takes the last coin from the center.

Variation:
As the children's skill increases, have every child take two coins from the center when a flipped coin lands "heads up." This will increase the number of coins to be divided evenly when a coin lands "tails up."

This activity also helps children learn about:
counting.